Problems – Ideas - Solutions

**Accidents-free landing of rockets.
An International Virtual League for Solution of Accidents-Free Landing Rocket Problems.**

Georgiy Tyshko

Problems – Ideas - Solutions

How Do We, People on Earth, Want to Live Tomorrow?

What kind of life would we like to see around us?
Making the life around us better, more comfortable, enjoyable, and effective in every way depends on us, on each and every one of us.

Let's build our future ourselves

Book Five – Accident-free landing of rockets. An International Virtual League for Solution of Accident-Free Landing Rockets problems.

Preface 5

Part One - Analysis of the problems of accident-free landing stages of rockets 8

Part Two - Concept of solving the problem of accident-free landing stages of rockets 12

Part Three — An International Virtual League for Solution of Accidents Free Landing Rocket Problems 17

Part Four - Description of some activities in virtual realities of An International Virtual League for Solution of Accidents Free Landing Rocket Problems 22

    Appearance of baskets 22

    Materials and design calculations of baskets 24

    Activity of mathematicians, programmers and analysts 25

    Advanced solutions for ground infrastructure 26

    Devices for electromagnetic levitation 26

    Development of new ways of aircraft takeoff and landing 28

    Development of new types of aircraft for ultra-long and ultra-high-speed flights 30

    Development of technology for mass flights between Earth and Moon 31

    Activities of recruitment specialists. 34

    Activities of lawyers, managers, investors and other specialists 34

Part Five - Maximum development of virtual solutions 36

Part Six - How to get the results of the implementation of the concept of solving the problem of accident-free landing stages of rockets quickly and profitably 40

Afterword 51

Preface

Everyone knows the names of the leaders in the information technology industry.
Sergey Brin, Larry Page, Jeff Bezos, Mark Zuckerberg, Jack Ma, Arkady Volozh, Yury Milman and others.
All of them are certainly highly outstanding personalities.

However, Elon Musk in my opinion the most outstanding of the most outstanding.
Elon Musk was the first to understand the uniqueness of the genius Nikola Tesla.
Elon Musk returned to mankind the name and ideas of Nikola Tesla.
Elon Musk was the first to realize that it was time to change hydrocarbon energy to electric energy.
Elon Musk was the first to start a new era in space exploration
Elon Musk plans to completely change the life of all mankind on the planet Earth.

In particular, Elon Musk plans to provide high-quality and affordable Internet anywhere in the world. This means that at any place of another cosmic body too. On Mars or Moon.
Elon Musk is the closest to solving the problem of mining on the moon.
Elon Musk is exceptionally talented and stunningly hard-working.
However Elon Musk not God, and man and he one not can solve all problems that face humanity.

For example, one of the problems that hinders the creation of the global Internet is the problem of trouble-free landing of rocket stages.
In order to ensure the operation of a large number of satellites, it is necessary to safely return the rocket stages for re-use.

In the book I present to the reader, I describe a new approach to solving the problem of trouble-free landing stages of missiles.

The approach is really very new and very unusual and at the same time very natural and logical.

If Elon Musk will try to solve all the problems only by his company SpaceX, in order to fully explore all the options and all the possibilities of the new approach may take years.

However, the necessary solutions can be implemented within a few months, if you apply another unusual solution.

The solution is to mobilize the intellectual potential of the entire population of the planet Earth.

In this case, it will be possible in a very short time to implement not only a trouble-free landing stages of missiles, but also to implement the launch of missiles in a more economical way.

Moreover, the new approach can be successfully implemented for take-off and landing of aircraft as well as for the creation of a new type of aircraft.
The new approach will significantly accelerate the development of the moon and Mars.

The main proposed solution is to create An International Virtual League for Solution of Accidents Free Landing Rocket Problems

Discussion of the proposed solutions is available at
howwewanttolive.livejournal.com
tag = Landing
tag = Rocket

All informative comments will be published

All relevant questions will be answered

Part One - Analysis of the problems of accident-free landing stages of rockets

We will conduct a small analysis of the processes that occur during the launch of rockets

At the time of launch, the rocket is always in a stable vertical position on the launch pad.

Note for the future that the rocket launches are made with minimal wind precisely because of the need to ensure a stable vertical position at the start and in the first seconds after the start

In order for the rocket has had minimal resistance in flight, the engines are located as closed as possible to the main body of the rocket

In order for the rocket has had minimal resistance in flight, the supporting "legs" are located as closed as possible to the main body of the rocket

This arrangement of engines and support "legs" makes the rocket vulnerable to crosswind on the launch pad

This arrangement of engines and support "legs" makes the rocket vulnerable to any lateral displacement in the first seconds of the flight

If the engines are located as close as possible to the main body of the rocket then in order to compensate for minor deviations from the vertical position of the rocket requires considerable effort and over operational control of the engines

The more time passes from the start of the rocket, the farther the rocket is removed from the surface of the earth and the higher the speed of the rocket flight becomes the less problems with the control of its position in flight.

Any pitch angles, any yaw angles and any roll angles can be returned to the required position. The surface of the earth is far away and the deviation of the rocket from the required position can always be compensated, without fear of falling to the surface of the earth

Let us further conduct a small analysis of the processes occurring during the landing stages of missiles

If during the launch of the rocket you can always wait for the weather with a little wind or with a complete lack of wind, then it is much more difficult to have the necessary windless weather at the time of the rocket landing

Of course, no one would think to launch and land missiles during a storm, but in any case, the wind during the landing of missiles will always be on average stronger than during the launch of missiles

Therefore, it is always more difficult to hold the rocket in the required vertical position when landing the rocket than when starting the rocket

During the landing rockets of the requirements for the location of the engines opposite.

During the landing rockets the farther the engines are apart from the main body of the rocket the easier it is to control the vertical position of the rocket.

During the landing of the rocket, the wider the support "legs" of the rocket are placed, the more stable it is upon landing

Thus, there are conflicting requirements for the location of engines and support "legs"

At the time of start than engines and legs closer to the body of the rocket the better, and during the landing opposite than the engines and legs located further away from the rocket body the better

There are serious concerns that the current procedure for the return of missiles for re-use cannot, in principle, be implemented for the mass and safe return of missiles to Earth.

Unfortunately, the landing rockets for reuse has another unpleasant circumstance

For the return and landing of the rocket, a significant additional amount of fuel is required, which must be on Board the rocket during the flight

Pay attention also to the fact that the flame that escapes from the nozzle of the engines, creates a total hell.

If during the launch the rocket very quickly leaves the domain of this utter hell, during landing, the rocket falls into this area and it continues to be there some time

As a result, there are serious concerns that mass launches are not possible under the existing missile landing procedure

At first glance, it seems that the contradictions to the requirements for the location of engines and support legs are irreparable

However, let's not panic because there is a solution

Part Two - Concept of solving the problem of accident-free landing stages of rockets

We will continue our analysis and build the necessary solution.

We would like when the rocket on those parts of the trajectory on which it has a high speed, the engines were located as close as possible to the main body of the rocket
At the same time, during take-off and especially during landing, we would like the engines and support legs to be as far away from the main body of the rocket as possible

In order to resolve this apparent contradiction, I propose that we proceed as follows

No one is surprised by the fact that the missiles have several stages

I suggest adding another stage of rocket, but a bit unusual

The added stage of rocket is something like a round basket

The diameter of the basket is much larger than the diameter of the main body of the rocket

Missile and / or screw engines are installed along the perimeter of the basket

The main body of the rocket in its lower part is located in the basket

Consider the process of take-off

The engines of the basket to raise the entire structure to a predetermined height

At this height, the engines of the main body of the rocket start

After the start of the engines of the main body of the rocket, the basket and the main body of the rocket are undocked

The main body of the rocket rises to a new predetermined height while the basket carries out the landing

The basket has a large diameter.
All engines are located on the outer circumference
Landing such a basket is no problem
Moreover, the larger the diameter of the basket the better in all respects

The larger the diameter of the basket the less likely damage to the basket by engines of the main body of the rocket
The larger the diameter of the basket the greater the reliability of a trouble-free landing of the basket in almost any weather

Consider the process of landing

Before landing, the main body of the rocket or its stage is lowered to a predetermined height under the control of its own engines

At the same height rises the corresponding basket under the control of its own engines

On this predetermined height, the rocket or its stage is smoothly joined with the basket

At the beginning of the landing process, the basket moves up towards the returned rocket stages

But at some point the basket stops and begins to move down

Thus the stage of the rocket gradually catching up with the basket when moving down

Thus, the process of combining the devices for docking the basket and the rocket stages is slow and easy to control

In the end, there is a docking basket and stages of the rocket

After docking, the landing control is carried out by the basket engines

Let's consider what advantages the proposed solution has.

First, the launch of the missile is possible in a very wide weather range.
The large diameter of the basket allows you to launch with a significant wind.

Second, the rocket can take a much larger payload because there is no need to carry the fuel necessary for landing at ends of the flight.

Third, during docking, the main body of the rocket and the basket will be maneuvered simultaneously, which greatly simplifies and facilitates the docking process, unlike docking with a fixed platform

Fourth, landing the whole structure (basket plus the main body of rocket) may be on any area.

On the roof of a skyscraper or on the site in the city center.
For example in Central Park :)
The fact that the engines of basket can be fully or partially screw motors

Fifth, take-off and landing places can be combined in one place

Sixth, the maintenance infrastructure of the landed missiles can be located in the same place of landing structures (basket plus the main body of rocket)

Thus, checking the technical condition of the rocket stages, dismantling of unusable parts, installation of new or restored parts can always occur automatically in the same place after landing

Placement of cargo sent to the near-earth space will also occur automatically at the landing site.

This is the place to start will always be located fully ready for take-off (basket plus the main body of rocket)
If the complex for some reason will be deemed unsuitable for the next take-off, it will be dismantled or will fly to the repair site
Each suitable for take-off complex will return to the place of take-off and landing on their own

At this point, the reader will exclaim emotionally.
What you described above may really solve the problem of accident-free landing of rocket stages.

But the existing technology of missile flight was created more than a hundred years!

There is a huge number of questions that need to be answered in order to implement something similar to what you have described!!!
What should be the diameter of the baskets?
Which engines should be placed in the basket?
At what height to carry out docking basket and rocket stages?

And a huge number of other questions that need to be answered!

In order to realize the idea you propose, it will take at least another hundred years!

For this reason all that you offer is a beautiful fairy tale but nothing more!

The reader is right at 99.99%
But not 100%.
The fact is that all the questions can be answered in a few months.

How exactly can be quickly answered all the questions necessary for the implementation of the proposed solution in practice, will be described in Chapter Part Three - An International Virtual League for Solution of Accidents Free Landing Rocket Problems

Part Three — An International Virtual League for Solution of Accidents Free Landing Rocket Problems

The reader is absolutely right.
In order to get answers to a huge number of questions that arise if you want to realize the take-offs and landings of complexes (basket plus the steps of the main body), it is necessary to use a huge scientific and technical potential.

Not even a large company in the World has such potential.
Even a group of large companies in the World does not have such potential.
Theoretically, the leading industrial States in the World could try to cope with this task.
But in practice, this is also unrealistic, primarily due to the inability to find the necessary specialists and organize joint work of these specialists.

However, all questions can be answered in a few months

It is necessary to create An International Virtual League for Solution of Accidents Free Landing Rocket Problems
.
Such a League should and will be a set of separate, independent Virtual Realities to Address the problems of Trouble-free landing Stages of Missiles.

The reader has the right to ask-why do you think that there are those who want to create such independent virtual Realities to solve these problems?

I explain

There are several reasons why such a League will appear.

The first reason — The exploration of near-earth and distant outer space came to a standstill
The further development of civilization requires regular and mass rocket flights.
In the existing technology of takeoffs and landings of missiles is impossible.
For this reason, all industrial countries will Finance creative activities aimed at solving the problems of regular and mass rocket launches.

Gaining the leadership in the area of regular and massive launches of missiles gives the leaders a huge commercial advantage
For this reason, most of the major corporations and their leaders will also Finance creative activities aimed at solving the problems of regular and mass rocket launches.

The second reason — The creation of virtual models of complexes (basket plus the main body stage) and modeling of takeoffs and landings of such complexes can be very profitable business due to the display of advertising. Similar to Disneyland, cable TV and YouTube.

The third reason - every person has a need to Express their opinion on any issue and the more there is a need to Express their opinion on the most important issues of reality.

Any person will always show curiosity about how other people offer to solve in this case the problem of Accident-free Take-off and Landing of Rockets.

Any person after getting acquainted with the proposals of other people will always want to Express his own ideas and opinions.

The only external impact that is necessary is to provide a comfortable place where the above reasons can start working.

As such a decentralized global forum to address the problems of Accident-Free Launching and Landing of Rockets, I propose the creation of An international Virtual League to Address the Problems of Accident-Free Launching and Landing of Missiles.

In order for the proposed international Virtual League to begin to be created, it is necessary to announce the necessity and possibility of its creation as much as possible.

Once the Information about the International Virtual League for Solution the problems of Accident-Free Launching and Landing of Rockets more or less spread in the World, the virtual realities included in it will begin to appear like mushrooms after the rain.

As a result, the collective intelligence of the planet Earth will start working.

This collective intelligence will begin to formulate questions on the way to the creation of the proposed complexes (basket plus stages of the main body of rocket)

For example.

How can the proposed complexes look like?
The basket will be similar to the dryer for the dishes consisting of strips or will be similar to the saucepan that is to have some solid body?
What sizes should be in baskets complexes?
What materials can be used to make baskets of complexes?
How many and what engines can be installed on the baskets of the complexes?
At what height should the docking and undocking of baskets and main rocket hulls be carried out?
Can the baskets of the complexes be operated by the crew?
Baskets of complexes can be fully automatic ?
The degree of profitability of operation of the proposed complexes?
What are the technological advantages of operation of the proposed complexes?

The questions will be well reasoned and convincing.

However, the same collective intelligence will always find solutions that will provide answers to emerging questions.

This book is written in order to give the first impetus to create An International Virtual League for Solution of Accidents Free Landing Rocket Problems (IVLSAFLRP)

Virtual reality members of IVLSAFLRP can be created on different platforms. Namely on the platforms Second Life, Sansar, Decentraland and platforms of other information technology leaders.

It is natural to expect that all members will be linked by links to move between virtual realities.

The creation of a decentralized virtual League will lead to the development of more or less identical interfaces for communication within virtual realities.

As a result, the best solutions for creating virtual realities will become standards.

Part Four - Description of some activities in virtual realities of An International Virtual League for Solution of Accidents Free Landing Rocket Problems

We will describe briefly what topics and what areas are possible for discussion in virtual reality, which will be members of **An International Virtual League for Solution of Accidents Free Landing Rocket Problems (IVLSAFLRP)**

Appearance of baskets

Design companies, individual designers and ordinary people, especially young people will be happy to start creating virtual projects of the Complexes (basket plus the main body) Naturally, first of all, virtual projects of baskets will be created.

The entire collective intelligence of the planet Earth will take part in the creation of virtual projects of the Complexes. Among the virtual projects of the complexes there will be successful projects and will be unsuccessful, but there will definitely be tremendous finds.

Such projects can be tied to specific places of take-off and landing.
Virtual projects can and will include a sufficiently detailed infrastructure of cargo and fuel storage.
Such projects may include repair service infrastructure and so on.

Of course, such activities will be directed and supported by interested companies (such as SpaceX) and government organizations (NASA, RosCosmos and others)

Such projects will be an excellent opportunity for design teams and individual designers to Express their ideas and secure advertising

Such projects can be created on the orders of companies (SpaceX) and government organizations (NASA, RosCosmos and others)

The cost of creating such projects is extremely small because all these projects will be created in virtual reality

All virtual reality will be available for visitors to visit avatars, who will leave their opinions and comments

Avatars of project creators can be present in these virtual realities

Virtual reality, focused on design solutions may well be self-supporting due to the display of advertising and data collection about consumers

To start the process described above, you only need to create some virtual reality.

There can be many virtual realities focused on design solutions. In different countries and on different continents.

Some of them will not be popular, and some will get more fame and recognition

Materials and design calculations of baskets

Companies producing materials that are necessary to create baskets, with great pleasure will begin to advertise their products and their technologies that can be used in the creation of baskets.

Similarly, the companies for the production of baskets with great pleasure will begin to advertise and their technology, which can be used in the manufacture of baskets.

Similarly, the companies on production of engines for baskets with great pleasure will start to advertise the engines, which can be used in the manufacture of baskets.

Virtual realities focused on the manufacture of baskets can be a lot.
In different countries and on different continents.

Such realities will become for companies a permanent exhibition of their achievements.
In such exhibitions there will always be a lot of avatars of visitors.
In such exhibitions avatars of manufacturers and avatars of customers can hold meetings and make deals not only on the range of Complexes (basket plus the main body of rocket)

The companies and professionals listed above will be able to communicate with each other in their respective virtual realities. Such an opportunity will allow these companies to formulate General requirements for robotic machines for Assembly, disassembly, repair of Complexes (basket plus the main body steps).
The same opportunity will allow these companies to formulate General requirements for robotic machines for cargo placement

and refueling of Complexes (basket plus the main body of rocket).

The above companies and specialists will be able to develop General requirements for the means of monitoring the condition of the Complexes (basket plus the main body of rocket).

The companies and specialists listed above will be able to create joint virtual projects

This practice of joint projects will enrich all participants and improve the quality of the created Complexes (basket plus the main body of rocket)

Activity of mathematicians, programmers and analysts

The avatars of mathematicians and programmers analysts from around the World will offer, discuss and simulate various algorithms for approaching and docking baskets and stages of the main body of missiles

Avatars of visitors will be able to set any behavior parameters for such models.
For example, a side wind or emergency situations like the failure of some of the engines

Avatars of visitors will be able to leave their comments and ask any questions

The corresponding virtual reality will be international and therefore in these virtual reality will always be present avatars of any specialists

I have no doubt in the high professionalism of Elon Musk's staff. They have already solved a huge number of problems associated with the flight of missiles and landing missiles.
But the proposed solution of docking above the ground basket and the main body will require a lot of time because of the abundance of options
Therefore, I propose to conduct a massive public search, and specific final decisions of course Elon Musk can patent and keep secret

However, the number of original ideas and solutions will increase in any case
And the number of great professionals that Elon Musk wants to invite to work for him will also increase

Advanced solutions for ground infrastructure

Smooth and precise landing of the baskets together with the rocket stages will allow to make a landing Complexes exactly on some platform.
Such a platform can go down under the ground to the required depth.
The entire Complex (basket plus rocket stages) can then be moved to any underground Storage location for storage, maintenance and loading for the next flight
All movements should be carried out by devices based on electromagnetic levitation

Devices for electromagnetic levitation

In Storage facilities for Complexes (basket plus rocket stages) it is proposed to use electromagnetic levitation devices as the only means of moving any objects

The complexes are proposed to be moved on platforms.
It is possible by turning 90 degrees in a horizontal position
People and goods are offered to be moved in cabins similar to the cabins of big van.

Both the platform and the cab proposes to attach to the brackets. Other ends of such brackets will move inside the boxes, which will contain electromagnets

Currently, there are practically no companies that specialize in the production of devices for electromagnetic levitation

However, it is the emergence of virtual realities in which such devices will appear in the form of virtual prototypes that will ensure the development of the industry of such devices in real conditions

Virtual realities oriented devices for electromagnetic levitation will begin to appear in different countries and on different continents.

Such realities will become for companies a permanent exhibition of their achievements.
In such exhibitions there will always be a lot of avatars of visitors.
In such exhibitions avatars of manufacturers and avatars of customers can hold meetings and to bargain not only for the nomenclature of Complexes (basket plus stage rocket), and not only the item's underground Storage Complexes

Development of new ways of aircraft takeoff and landing

The above proposed concept of take-off and landing stages of missiles can be extended to a similar concept of take-off and landing aircraft

Instead of a round basket in this case, it is proposed to use a "bed" that follows the contours of the aircraft

In such a "bed" is a sufficient number of jet and / or screw engines.

The "bed" also has a large number of powerful chassis, which are located far from each other

Such the "bed" can independently take off and land

Such the "bed" can take off and land on a very short strip or can take off and land vertical

At the time of take-off, the aircraft engines and the "bed" engines provide take-off either from a short strip or vertically.

Takeoff occurs with the rapid climb

This rise is reminiscent of the take-off and climb of military aircraft with extra fuel tanks with fuel.

The role of additional tanks in this case plays the proposed "bed".

The "bed" is a powerful plane, which provides a quick lifting of the plane and the rapid descent of the aircraft.

Docking and undocking of the aircraft and the "bed" occur at a given height.
This ensures a quick and safe take-off and a quick and safe landing.

The "bed" will have brackets, the ends of which will move in boxes on the principle of magnetic levitation

Thus, the use of the "bed" will completely eliminate the process of taxiing before take-off and after landing

The use of the proposed "bed" will significantly reduce the take-off and landing time
The use of the proposed "bed" will increase the weight of the cargo

The use of the proposed "bed" involves the use of conventional aircraft

Such aircraft can take off and land in the same way as they always did

In order for the aircraft to take off and land using the bed, it is necessary to equip the aircraft with the means of docking and undocking control of the aircraft and the "bed"

The reader may ask

Why not immediately study the details of the creation of Systems for take - off and landing missiles at the same time with the proposed "bed"?

In my opinion, this should not be done for the following reason

Complexes for safe take - off and landing of rockets are necessary at the moment "as air"
The possibility of creating such Complexes is obvious
And such complexes can be developed and created within months.

At the same time, the proposed "bed" described above is a draft project, all the pros and cons, which will first be studied

Therefore, the feasibility of the proposed a Bed should first be studied
While Complexes of take-off and landing rockets should start to create

These are two different activities although very similar in theme

Development of new types of aircraft for ultra-long and ultra-high-speed flights

The concepts proposed above for the take-off and landing of rocket stages and the take-off and landing of aircraft can be extended to the development of new types of aircraft
The proposed new type of aircraft is a Complex consisting of a basket or "bed" on one side and the main body

The body of such an aircraft is something in between the body of the rocket and the body of the aircraft
This body is a body of the aircraft designed to carry goods and passengers over long distances and at high speeds
On the other hand, such a body is a rocket because it is completely devoid of wheels, that is, it can not take off and land on its own

The engines of the proposed new aircraft will also represent a cross between the well-known rocket engines and turbo fan engines, much larger than the existing ones

The proposed aircraft of the new type will have significantly more swept wings convenient to maintain significantly higher speeds in the upper atmosphere but completely unsuitable for landing

The proposed aircraft will not be widely used even if created.

But the need for such devices certainly exists.

Therefore, consideration of aspects of the development of such devices operation makes sense to include in the area of interests of the proposed virtual reality An International Virtual League for Solution of Accidents Free Landing Rocket Problems (IVLSAFLRP)

However, the development of aircraft of the new type proposed above should be considered as a continuation of the development of the concept of creating aircraft from separable full-fledged parts

While Complexes of take-off and landing rockets should create right now

Development of technology for mass flights between Earth and Moon

The above-proposed concept of take-off and landing of stages of rockets can be used to organize mass flights between the Earth and the Moon.

In this case, there are two possible solutions to the problem of mass flights

In the first case, the Complex (basket plus the main body) during take-off from the Ground rises to a sufficiently high altitude, where the separation of the basket and the main body.
After separation, the main body is directed to the Moon
Towards the main body flying up to the Moon, the lunar basket takes off from the surface of the moon
At a sufficiently high altitude there is a soft docking of the lunar basket and the main body
Moon basket provides a smooth landing of the whole Complex
After landing, the lunar basket moves along the magnetic levitation boxes to the technical sublunary premises

The process of flight from the Moon to the Earth takes place in reverse order

In the second case, the Complex (basket plus main body) during takeoff from the Ground rise to the height of the near-earth orbital station

Further, two sub-cases are possible

In the first sub-case the main body is located in a special very large module designed only for flights between near-earth and near-moon spaces

In this special module can be placed several main bodies of rockets

After filling such a module with the main rocket hulls, such a module begins to fly from the Earth to the moon or back

In the second sub-case the main body is joined to the near-earth orbital station, to which containers with cargo and passengers are moved by magnetic levitation devices

In the future, containers with cargo and passengers are moved to a special module for the flight from the Earth to the Moon

In the orbit of the moon is the lunar orbital station, through which containers with cargo and passengers are moved to the main buildings of lunar missiles.

Lunar missiles do not need to use additional steps in the form of baskets for landing
On the Moon there is no atmosphere and the force of attraction is six times less.

Due to the lack of atmosphere, the engines and support legs can be located far enough away from the main body of the rocket
Due to the small force of gravity on the moon, fuel savings due to the use of pots will be minimal while the complexity of docking and undocking pots and main rocket hulls will remain the same.

The use of special modules for the flight from the Earth's orbit to the orbit of the Moon and back will provide the necessary protection of passengers from hard radiation
The body of such modules can be thick enough and can be made of lead.

The development of the concept of mass flights from the earth Moon should be considered as a continuation of the development of the concept of creating aircraft from the separated full-fledged parts.

But the final decisions on the implementation of the concept should be made only after a full feasibility study of the need for such flights

Elaboration of the above-described concept will be used to carry out technical and economic feasibility of such flights.

This is an important but still perspective

While Complexes of take-off and landing rockets should create right now.

Activities of recruitment specialists.

Among the virtual realities included in the **An International Virtual League for Solution of Accidents Free Landing Rocket Problems (IVLSAFLRP)** will necessarily appear virtual reality, which will specialize in recruitment
In these same realities avatars of specialists will be able to tell about themselves much more than they can tell about themselves with the help of resume and CV.

Activities of lawyers, managers, investors and other specialists

The reader has probably already guessed that any activity necessary to create Complexes (basket plus rocket stages) and underground Storage for Complexes will be presented in specialized or mixed virtual realities

Avatars of lawyers, managers , investors and other professionals will be able to gather and communicate at any time, no matter in what countries of the World they are, and not only in what cities.

Such meetings can be of different degrees of openness.
Open meetings and records open meetings can be accessible to any visitors of the avatars.

In such virtual realities can be discussed and concluded any financial transactions and any contracts

In such virtual realities can be selected the necessary staff and much more

It should be noted that all of the above and the above activities in the previous paragraphs are virtual

However, more on this in the next chapter.

Part Five - Maximum development of virtual solutions

I accept the idea that everything that has been read so far seems highly unusual to the reader. Moreover, I admit that the reader as they say his head is spinning.
Where to start and what to do?

I will try to help the reader to better understand the processes of obtaining the necessary solutions.

First. All the books in the series "Problems — Ideas — Solutions" include the formulation of problems, generation of ideas and the formation of solutions in Virtual Reality

Second. The optimal functioning of each virtual reality is possible only within the framework of international virtual leagues uniting virtual realities, which are aimed at finding solutions to relevant problems

Third. Formulation of problems, generation of ideas and formation of solutions in virtual reality will mobilize the huge intellectual potential of the World to find solutions to any problems. Any person in the form of his avatar will be able to take full part in the work of any virtual reality.

Fourth. The most important and most important rule or principle, call it what you want, is the rule or principle of the maximum possible implementation of any solution in the corresponding virtual reality.

This means literally the following — any object must always first be created in the corresponding virtual reality.

It's hard to understand the first time.
So I'll try to explain with examples.
For example, you need to build a bridge.
In accordance with the proposed rule or principle, the bridge should first be built in virtual reality and best of all not in one, but in several independent virtual realities.

Any object is thousands of times easier, faster and cheaper to create in virtual reality.

Once the object is created in virtual reality to evaluate all the pros and cons of this object can be a huge number of people from around the world

You can be absolutely sure that among these people will be much more intelligent people and very good professionals than in any team that can currently be hired to create a bridge.
Even if this team has geniuses like Robert Moses.

Once the object is created in virtual reality, you can easily create models for the operation of this object.

In 2017-2018 in new York, in Brooklyn on the Belt Pkwy was built car overpass between 13-m and 9-m exits.

Starting from the 13th exit to the East the Belt gives and takes cars. Starting from the 9th exit to the West the Belt also gives and takes cars.
Between the 9th and 13th exits, the number of vehicles on the Belt is always greater than in other sections

Between the 9th and 13th exits on both sides of the Belt is an empty and contaminated area.

Consequently, nothing prevented to do in the overpass between the 9th and 13th exits four lanes in each direction.
However, this has not been done.

I do not insist that this place should have four lanes in each direction.
Must have been some reason to leave three lanes in each direction.

But I am sure that quite a lot of people would pay attention to the insufficient, in their opinion, the number of bands between the 9th and 13th exits
Moreover, independent mathematical models of the behavior of this area would objectively demonstrate the influence of the number of lanes in this area on the speed of movement in this area.

Even if someone and there were thoughts about the need for more lanes in this area, such a person was not able to Express his doubts and discuss these doubts with other citizens who thought as well

It is not necessary to engage in demagoguery and say that such people could apply to the city administration or write to some newspaper.
It is clear to everyone that there should be a standard procedure by which important public decisions can be discussed widely before such decisions are actually implemented.

Even if such well-developed processes as the construction of an overpass can cause questions and discussions, absolutely new and unusual processes, such as the development of Complexes (basket plus rocket stages) for a trouble-free landing of rocket stages, especially should first be created and discussed in virtual reality.

The creation of Complexes (basket plus rocket stages) for trouble-free landing rocket stages, at first in virtual realities will allow us to fully simulate and explore all aspects of such systems in reality

How to speed up the process of implementation of the proposed concept and how to make this process as profitable as possible for all participants will be described in the next paragraph.

Part Six - How to get the results of the implementation of the concept of solving the problem of accident-free landing stages of rockets quickly and profitably

Quite a lot of earthlings feel sympathy and respect for Elon Musk.
A very large number of people are anxiously watching the successes and failures of companies run by Elon Musk
Even more people would like to accelerate the movement in the directions that he started.
The development of near-earth and remote space is one of these areas

One of the problems that stand in the way of accelerated development of near and far outer space is the mass accident-free landing of missiles.

Above I described the Concept of solving the problem of accident-free landing stages of missiles.
The solution can be achieved by creating An International Virtual League for Solution of Accident-Free Free Landing Rocket Problems (IVLSAFLRP)

Probably everyone who got acquainted with this book will tell you.
Yes, you are right to ensure trouble-free landing of missiles in large numbers is currently impossible.
If something similar to what you suggest was possible to implement, the takeoffs and landings of missiles would be even easier than the takeoffs and landings of aircraft at the moment.

But it is not clear how such a Concept can be implemented in the foreseeable future.
How can ordinary people influence the process of moving towards solving the problems you have described?

In this paragraph I will answer this question.
This paragraph is addressed to everyone without exception. That is, to every inhabitant of the Earth.

The first question I ask you to ask yourself.

Do you agree that there are big problems with mass flights and missile landings?
And do you agree that there are big problems with the ultra-fast delivery of people and goods between the continents?
And do you agree that in the future there are big problems with the delivery of people and goods between the Earth and the Moon?
Answer Yes or No.

If you think that there are no problems with mass flights and missile landings, you can safely close this book

However, if you answered Yes, please answer the following question
You want to the situation of mass flight and landing of the rockets really changed and as soon as possible?
Answer Yes or No
Perhaps you agree that the problems exist but they do not really interfere with you and you do not want to break your head about it.
No one will judge you, there are enough other problems in life to think about them.

However, if you agree that the problems with mass flights and missile landings exist and if you want these problems to be solved as soon as possible, then I ask you to do the following

First, ask all of your friends to answer the two questions I asked you to answer

Leave alone those who answer at least one question - No

Those who are willing to contribute to the solution of problems with mass flights and landings of missiles, and those who are willing to contribute to the solution of problems with the super fast delivery of people and goods between the continents ask to do the same thing that I ask you to do

Second, tell these people something like this
You know-I read about one idea of solving the problem of mass flights and landings of rockets and solving the problem of super fast delivery of people and cargo between continents and mass flights to the moon.

These problems can be solved completely and in a very short time.

This idea involves the creation of virtual realities in which the avatars of the participants will discuss specific solutions and visually observe the virtual models offered.

Do you know what virtual reality Second Life, Sansar, Decentraland and virtual realities of other information technology leaders are?

If these people do not know and have never heard, then try to explain to them and advise them to go to these virtual realities

Then you tell these people the following.

Imagine that in these virtual realities there are places where you can create virtual projects of Complexes in action (basket plus main body of rocket) for trouble-free takeoff and landing of missiles.

Your interlocutor quite possibly will ask you, and how to find them there?
You can honestly say that you do not know.

Then you can ask yourself and your interlocutor.
Maybe you should tell other people about this idea?
About the idea of addressing trouble-free takeoff and landing of missiles through the creation of An International Virtual League for Solution Free of Accidents Landing Rocket Problems (IVLSAFLRP)

Among the people you can tell about this idea may be people who constantly have to deal with the delivery of people and goods, journalists and people who are able to develop the proposed virtual realities.

You and your interlocutor are likely will agree.
Yes, there are interested businessmen, journalists and specialists. But why would they suddenly agree to take some action towards the creation of the described virtual realities?

I will explain why all of the above will enthusiastically act towards the creation of the described realities.

Interested businessmen.

Any businessman, whose business is related to the delivery of people and goods between the continents, will always be interested in the latest developments in this area

The cost of creating small local virtual realities, in which the avatars of visitors will be able to discuss options for creating Complexes (basket plus main body of rocket) necessary for each specific business will be minimal.
Namely, the cost of such developments will be no more than the cost of creating a small site.

As soon as a few of these simple virtual realities, they will unite in an International League dedicated to the creation of unified Complexes (basket plus main body of rocket).

The number of such virtual realities will begin to grow.
Between these virtual realities will begin in some way a competition about who will be better to cover the problem and who will be better to implement
virtual Complexes (basket plus main body of rocket).

As a result, the world wide web will be a place where there will be a collective brain aimed at solving the problems of creating unified Complexes (basket plus main body of rocket).

Press people.
A journalist of any country and any media is always interested in writing on topics that interest most people.
The solution of problems with mass flights and landings of missiles and problems with the ultra-fast delivery of people and goods between continents is certainly a topic that interests a large number of people in all countries of the World.

Specialists in the creation of virtual realities.

At present, great progress has been made in creating virtual realities of various types and levels of complexity.

However, at the same time, the contents of virtual realities mark time in the same place.

A huge number of virtual realities are fictional worlds in which monsters fight and half-naked beauty

In other virtual realities reproduced battles between soldiers of different countries and peoples at different times.

A large number of virtual realities create the supposed world of other planets.

The most powerful virtual reality Second Life was created in 2003 but until now it is focused only on young people up to about 30 years

In other words, many specialists in the creation of virtual realities from large teams to individual specialists will be happy to work in the new field of creating virtual realities for the creation of Complexes (basket plus main body of rocket).

Namely, Complexes that will solve problems with mass flights and landings of rockets and problems with super fast delivery of people and goods between continents and between the Earth and the Moon

I have listed above the selfless motives that will attract leaders, journalists and specialists in the field of virtual reality development to create Complexes (basket plus main body of rocket)..

But the proposed virtual reality can bring significant profits to all participants even before the moment when the Complexes (basket plus main body of rocket) will begin to successfully take off and land

The fact is that any virtual reality are place where advertising can be placed

With regard to advertising, virtual reality will have many advantages over existing opportunities to deliver advertising at sporting events, through youTube, Google, Facebook, Cablevision and other media

Thus, the more interesting will be for visitors avatars virtual reality dedicated to solving the problems of creating Complexes (basket plus main body of rocket) the more funds can be obtained for the development and operation of the virtual reality

As a result, there will be a large number of virtual realities focused on solving the problems of creating Complexes (basket plus main body of rocket).
Most of these virtual realities will be profitable even before the mass of successful takeoffs and landings of Complexes through advertising.

As a result, the virtual realities of An International Virtual League for Solution of Accidents-Free Landing Rocket Problems (IVLSAFLRP) will begin to appear quite well-designed virtual projects Complexes for trouble-free takeoffs and landing of missiles and to move people and cargo over very long distances As well as related infrastructure projects

From this moment on, any visitor to the avatar will be able to simulate various situations that may arise during the take-offs and landings of the Complexes (basket plus main body of rocket) and the functioning of the associated infrastructure

Accordingly, the advantages and disadvantages in the operation of such Systems for accident-free take-off and landing of missiles and for the movement of people and goods over very long distances will begin to be revealed.

Thus, the advantages and disadvantages will be identified before the start of operation of the Complexes (basket plus main body of rocket) in reality

At the same time, virtual projects of construction and installation works for the ground infrastructure of the Complexes (basket plus main body of rocket) will begin to appear)

The more visitors of avatars will visit virtual realities from An International Virtual League for Solution of Accident-Free Landing Rocket Problems (IVLSAFLRP), the greater will be the flow of advertising and therefore the greater will be the financial opportunities to improve and accelerate the development of virtual projects

The more visitors of avatars will visit virtual realities from An International Virtual League for Solution of Accident-Free Landing Rocket Problems (IVLSAFLRP), the greater will be the level of collective intelligence in the development of virtual projects, the higher will be the quality of decisions in the development of virtual projects

Dear reader, it is for the reasons listed above that you will be able to see and almost live check how the Complexes (basket plus main body of rocket) will look and function for trouble-free takeoffs and landings of missiles and for the movement of people and cargo over very long distances, including mass flights between the Moon and the Earth

Dear reader, you will have access to all information about the cost of the Complexes and the cost of construction and installation of the necessary infrastructure, you will have access to all information about the cost of operating costs and any other information before the Complexes in reality.

As a result, dear reader, you will have comprehensive information about the Complexes (basket plus main body of rocket)

The whole process of developing projects of such Complexes will be a fascinating and useful show for all the inhabitants of the Earth

In this show you will be both a spectator and a participant at the same time.

Dear reader, you will be amazed how quickly and efficiently will happen all that is described above!

The trick is that all of the above will be created in virtual realities from An International Virtual League for Solution of Accident-Free Landing Rocket Problems (IVLSAFLRP)

Creating any objects in virtual reality is hundreds of thousands of times cheaper and faster than in reality

Complexes (basket plus main body of rocket) for accident-free takeoffs and landings of rockets and for movement of people and cargo over very long distances will be created collective intelligence around the World for a few months!

However, you, dear reader, are interested in the end result. Namely Complexes (basket plus main body of rocket) for accident-free takeoffs and landings of rockets and for movement of people and freights on very long distances which can be used in reality.

All that has been written above about the creation of Complexes (basket plus main body of rocket) for accident-free takeoffs and landings of missiles and to move people and goods over very long distances in virtual reality, it is true to create Complexes in reality

However, missile flights are always a great danger to the world. Mass flights carry many times greater danger to the world around them

Therefore, it is especially important to be able to have a comprehensive discussion of as many specialists as possible from different areas of real processes of movement of people and goods over long distances and between the Earth and the Moon

Let's consider the issue of availability of information on the development and operation of Complexes and the necessary infrastructure

On the one hand, the more open and accessible to everyone information about the creation and operation of the complexes then more attractive will be An International Virtual League for Solution of Accident-Free Free Landing Rocket Problems (IVLSAFLRP) for all other visitors

On the other hand, the practice of implementing any project implies privacy and confidentiality.

At first glance, it may seem that we are faced with some conflict of interest
And to resolve this conflict requires some kind of external regulation.

However, it is not. The degree of privacy and confidentiality will automatically be self-regulated.

The more open and accessible to all those wishing information on the establishment and operation of the complexes the greater the number of visitors avatars.
The greater the number of visitors avatars the higher the efficiency of advertising.
Additional funds received from the provision of advertising can significantly reduce the cost of the creation and operation of complexes.

At the same time, the greater the number avatars of visitors, the higher the quality of design and operational solutions

Dear reader if you are really interested in a complete and comprehensive solution to the problem of accident-free takeoffs and landings of missiles and to move people and goods over very long distances, then tell your friends and acquaintances about the solution that I propose in this book

Afterword

First of all, I ask my readers to forgive me for the clumsy language and for no less clumsy presentation of the material. My only excuse is that I want to involve as many people as possible in the discussion about the need and possibility of creating **An International Virtual League for Solution of Accidents Free Landing Rocket Problems (IVLSAFLRP)**

In conclusion, I want to focus on two very important circumstances of the present time, which are figuratively speaking «knocking on our doors», but the huge potential reserves of these circumstances, humanity has not yet fully realized

The first circumstance that knocks on the door :)

Thanks to the global spread of the Internet, almost every person on Earth can virtually be present anywhere and participate in any discussion of any problems, ideas and solutions

The second circumstance that knocks on the door :)

More than ten years there is a virtual reality Second Life. Now there are virtual realities such as Sansar, Decentraland and virtual realities from other information technology leaders.

The possibility of visual contact with a potential companions, contains a great potential for obtaining and transmitting information.

The Chinese rightly say that it is better to see one picture than to read a thousand pages.
It is the possibility of "almost physical" communication of groups of people incorporated in the idea of creating <u>An International Virtual League for Solution of Accidents Free Landing Rocket Problems</u>.

I am sure that the possibility of «almost physical» communication of people in virtual reality to solve any problems will provide a huge technical, technological, scientific, moral leap for humanity

In the series of books "Problems-Ideas-Solutions" I want to offer for General discussion the following topics

The problems of transport development
The problems of assistance in emergency situations
The problems of the deficit and surplus stocks
The problems of the most acute shortage of human communication
The problems of application of magnetic levitation actuators instead of the wheels
The problems of the use of virtual realities instead of web sites
The problems of creating collective intelligence
The problems of application of methods Of analysis of problems and Synthesis of Solutions
The problems of education in virtual reality
The problems of evaluation of creativity of individuals
The problems of teaching fundamental Sciences
The problems of system approach in solving any problems
The problems of creating virtual prototypes of any devices, ranging from simple household appliances and laboratory equipment and ending to giant structures on Earth and in Space

The problems of creating a Global Integrated Common Physical and Virtual reality

The problems of regular effective interaction between presidents and Prime Ministers on the creation of a new type of society on Earth

I will be very grateful to everyone who will provide material support to my plans

PayPal   for account#   georgiytyshko@yahoo.com

Georgiy Tyshko   October 2018

www.ingramcontent.com/pod-product-compliance
Lightning Source LLC
Chambersburg PA
CBHW030511220526
45464CB00006B/2751